SPOTLIGHT ON OUR FUTURE

PLASTIC POLLUTION AND OUR EARTH

GENE BROOKS

NEW YORK

Published in 2022 by The Rosen Publishing Group, Inc.
29 East 21st Street, New York, NY 10010

Copyright © 2022 by The Rosen Publishing Group, Inc.

All rights reserved. No part of this book may be reproduced in any form without permission in writing from the publisher, except by a reviewer.

First Edition

Editor: Theresa Emminizer
Book Design: Michael Flynn

Photo Credits: Cover FatCamera/iStock/Getty Images; (series background) jessicahyde/Shutterstock.com; p. 5 Maxim Blinkov/Shutterstock.com; p. 6 Apic/Hulton Archive/Getty Images; p. 7 Jeff Greenberg/Universal Images Group/Getty Images; p. Ekaterina43/Shutterstock.com; p. 9 https://en.wikipedia.org/wiki/Nylon_riots#/media/File:Standing_in_line_for_nylon_stockings_at_Millers_dept._store_Oak_Ridge_(7851145488).jpg; p. 11 (background) Alina G/Shutterstock.com; pp. 11, 13 (poster) courtesy of Library of Congress; p. 12 https://en.wikipedia.org/wiki/File:Richard_Nixon_presidential_portrait.jpg; p. 15 Jag_cz/Shutterstock.com; p. 16 Jekurantodistaja/Shutterstock.com; p. 17 Mohamed Abdulraheem/Shutterstock.com; p. 19 Bloomberg/Getty Images; p. 21 Allen J. Schaben/Los Angeles Times/Getty Images; p. 22 Justyna Sobesto/Shutterstock.com; p. 23 Andreas Rentz/Getty Images; p. 24 Michel Porro/Getty Images; p. 25 Robin Utrecht/AFP/Getty Images; p. 27 (both) Charlie Crowhurst/Getty Images; p. 29 Visharo/Shutterstock.com; p. 30 Tuomas Lehtinen/Shutterstock.com.

Cataloging-in-Publication Data

Names: Brooks, Gene.
Title: Plastic pollution and our Earth / Gene Brooks.
Description: New York : PowerKids Press, 2022. | Series: Spotlight on our future | Includes glossary and index.
Identifiers: ISBN 9781725324206 (pbk.) | ISBN 9781725324237 (library bound) | ISBN 9781725324213 (6 pack)
Subjects: LCSH: Plastics--Environmental aspects--Juvenile literature. | Plastic scrap--Environmental aspects--Juvenile literature. | Plastics--History--Juvenile literature.
Classification: LCC TD798.B766 2022 | DDC 363.738--dc23

Manufactured in the United States of America

Some of the images in this book illustrate individuals who are models. The depictions do not imply actual situations or events.

CPSIA Compliance Information: Batch #CSPK22. For further information contact Rosen Publishing, New York, New York at 1-800-237-9932.

CONTENTS

TOO MUCH PLASTIC 4
A NEW MATERIAL 6
HELPFUL PLASTIC 8
MOUNTAINS OF PLASTIC 10
RECYCLING PLASTIC 12
PLASTIC POLLUTION 14
BREAKING DOWN THE PROBLEM 16
MYTHS AND FACTS 18
HOW TO USE LESS PLASTIC 20
GOODBYE, PLASTIC BAGS 22
INVENTOR BOYAN SLAT 24
NEW MATERIALS, NEW ANSWERS 26
HOW YOU CAN HELP 28
A NEW WAY FORWARD 30
GLOSSARY 31
INDEX 32
PRIMARY SOURCE LIST 32
WEBSITES 32

CHAPTER ONE
TOO MUCH PLASTIC

If you look around at things such as toys, shampoo bottles, and cars, you might notice many are made of a **material** called plastic. Plastic is very useful. It's also very harmful. Plastic waste can hurt many plants and animals as well as planet Earth.

Plastic is a man-made material created from **fossil fuels** such as oil, coal, and natural gas. Scientists first invented plastic only a little over 100 years ago. Since then, people have created over 9.2 billion tons (8.3 mt) of plastic. Scientists have found plastic waste all over Earth, including in the Arctic.

Plastics can be recycled, or remade, into new items. Unfortunately, very little plastic gets recycled. Governments, groups, and people like you are working to solve the plastic problem and break our wasteful habits with plastics.

Humans waste too much plastic. About 8 million tons (7.3 million mt) of plastic gets into the oceans every year.

CHAPTER TWO

A NEW MATERIAL

Before plastics, people used wood, metal, cotton, glass, animal bones, and other materials to make things. Elephant **tusks** were used to make piano keys and other items. Many elephants died because of this, so people wanted to find something new to use.

In 1869, a scientist named John Wesley Hyatt invented a new material. He used a plant material to create a plastic called **celluloid**. Many things, including combs, buttons, toys, and dice, were made from celluloid.

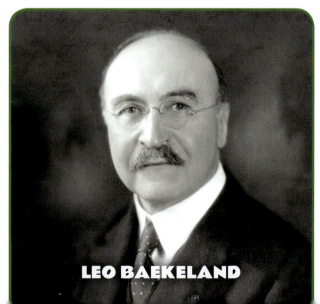
LEO BAEKELAND

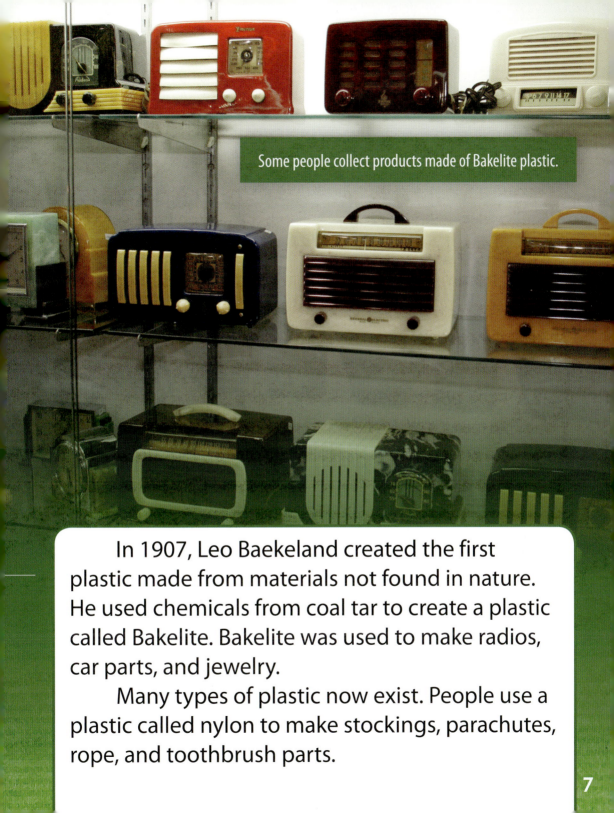

Some people collect products made of Bakelite plastic.

In 1907, Leo Baekeland created the first plastic made from materials not found in nature. He used chemicals from coal tar to create a plastic called Bakelite. Bakelite was used to make radios, car parts, and jewelry.

Many types of plastic now exist. People use a plastic called nylon to make stockings, parachutes, rope, and toothbrush parts.

CHAPTER THREE

HELPFUL PLASTIC

During World War II (1939 to 1945), many materials were in short supply. Plastic production grew by 300 percent. People made ropes from nylon and aircraft windows from plastic plexiglass.

Soon, companies used plastics to make items for the home. They made food containers, tables, and kitchen countertops from plastic.

Today, manufacturers use Velcro to connect many things. A Swiss engineer named George de Mestral invented it using plastic in 1941. He got the idea from plants that stuck to his clothes. Now, diapers and sneakers may use Velcro. Even astronauts in space use it.

These women are waiting in line to buy nylon stockings in 1946.

Plastics have helped make many things safer. Plastic can be light and strong, so manufacturers use it for many car parts and other things. The medical field uses plastic for face masks, gloves, and gowns.

CHAPTER FOUR

MOUNTAINS OF PLASTIC

In the 1960s, most people in the United States used about 30 pounds (13.6 kg) of plastic a year. Today, most people use about 220 pounds (99.8 kg) of plastic a year. In 2019, companies created 300 million tons (272.2 million mt) of plastic. By 2050, there could be as much as 12 billion tons (10.9 billion mt) of plastic in landfills.

By the 1960s, people started to notice all the plastic waste. In 1965, a scientist first recorded finding a plastic bag in the ocean.

As more plastic waste piled up, people began to take more notice. In time, people planned a special day about the **environment**. On April 22, 1970, they held the first **Earth Day** in the United States. People sometimes consider this the start of the modern environmental movement.

Japanese artist Yukihisa Isobe created the first Earth Day poster.

CHAPTER FIVE

RECYCLING PLASTIC

The more plastic Americans used, the more people noticed plastic trash. A group called Keep America Beautiful started in 1953. They helped teach people not to litter.

There were more environmental laws too. The U.S. Congress passed the Clean Air Act in 1963. In 1965, Congress added the Solid Waste Disposal Act. In 1970, President Richard Nixon signed an order to create the Environmental Protection Agency (EPA).

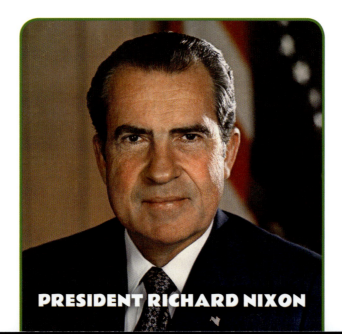

PRESIDENT RICHARD NIXON

Unfortunately, plastic pollution got worse. In 1969, scientists discovered that seabirds in Hawaii had eaten plastic by mistake.

To stop plastic waste, people started to reuse and recycle plastic. In the 1970s and 1980s, places around the United States started to recycle some plastic items. Even with recycling, though, plastic pollution kept getting worse.

This cartoon from the 1970s shows how many trash items can be recycled.

CHAPTER SIX

PLASTIC POLLUTION

About 71 percent of Earth is covered by water. A lot of plastic gets into the world's oceans each year. People who study plastic waste think between 8.8 million and 20 million tons (8 million and 18.1 million mt) of plastic winds up in the oceans every year.

An item that a person only uses once before throwing it away is called a single-use item. Plastic bottles, straws, and bags are all single-use items. About 40 percent of all plastic waste is single-use items.

Plastic has started to clump together in the oceans. In 1997, a sailor named Charles Moore discovered the **Great Pacific Garbage Patch** between Hawaii and California. Today, about 87,000 tons (79,000 mt) of plastic are in the Great Pacific Garbage Patch. Most of it is old fishing gear.

Plastic litter kills up to 100 million ocean animals every year. Leatherback sea turtles eat plastic bags by accident, thinking that they're jellyfish. This kills 100,000 of them every year.

CHAPTER SEVEN

BREAKING DOWN THE PROBLEM

Plastic trash on land is a problem too. Animals on land eat plastic by accident. They get caught in plastic string and stuck in plastic containers. Birds use plastic trash to make nests, which can be bad for baby birds.

Plastic is very durable. It breaks into pieces long before it biodegrades, or breaks down completely. No one is sure how long it takes plastic to biodegrade. Some think it could take as many as 450 years. Or, it may never biodegrade.

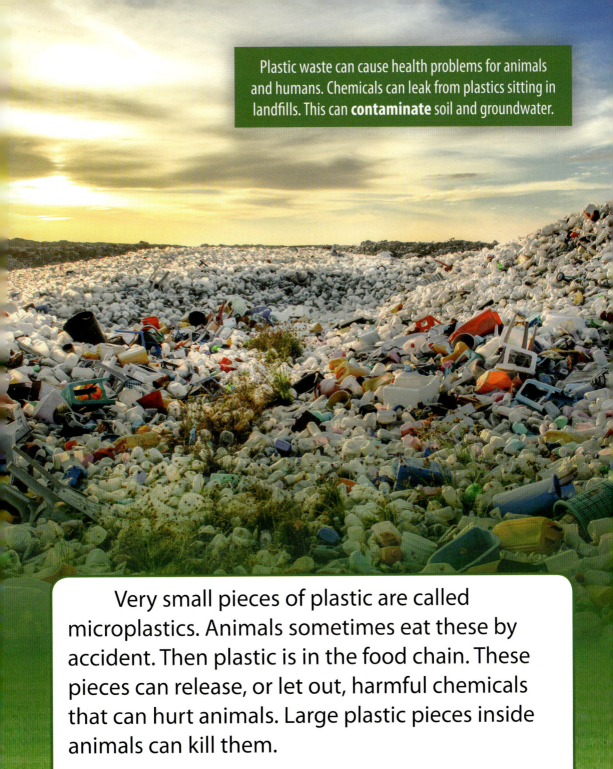

Plastic waste can cause health problems for animals and humans. Chemicals can leak from plastics sitting in landfills. This can **contaminate** soil and groundwater.

    Very small pieces of plastic are called microplastics. Animals sometimes eat these by accident. Then plastic is in the food chain. These pieces can release, or let out, harmful chemicals that can hurt animals. Large plastic pieces inside animals can kill them.

CHAPTER EIGHT
MYTHS AND FACTS

There are many myths about plastic. These are ideas people think are true but are really false. Here are a few common myths and facts about plastic.

Myth: Over time, plastic biodegrades and will be harmless to the environment.
Fact: Plastic is very sturdy. No one is exactly sure how long it takes to break down. Plastic created 100 years ago is still in the environment.

Myth: You can walk on the Great Pacific Garbage Patch.
Fact: The Great Pacific Garbage Patch isn't a floating island. Most of the floating plastic in garbage patches is hidden under the water's surface.

Myth: Not all plastic is recyclable.
Fact: Most plastic can be recycled. However, most plastics don't get recycled. Check to see if you can recycle plastics where you live.

You can't see all plastic pollution. A lot of the Great Pacific Garbage Patch is made up of small pieces of plastic called microplastics.

CHAPTER NINE

HOW TO USE LESS PLASTIC

Every plastic item you use adds up. Plastic pollution is still a growing problem. Since 2000, humans have used more plastic than in all the years before.

Hawaii and California were the first states to ban single-use plastic bags. Connecticut, Delaware, Maine, New York, Oregon, and Vermont have also banned single-use plastic bags.

You can start using less plastic now. Here are a few ideas to use less plastic and make a difference:

- Stop using plastic straws. Use a reusable straw instead.
- Stop using plastic cups when you buy ice cream. You can eat a cone and don't need a spoon.
- Use reusable water bottles instead of single-use bottles.

Reusable items can help you use less plastic. This way, everyone can help with the cleanup.

Make a difference! If you use reusable shopping bags all your life, you could keep 22,000 plastic bags from becoming trash.

CHAPTER TEN

GOODBYE, PLASTIC BAGS

Until 2017, China took in about half of all plastic collected for recycling. Then China stopped this practice to improve its own environment. Many countries didn't know what to do with their plastic recycling. In 2018, the European Parliament, or lawmaking body, voted to ban single-use plastic items. Europe also hopes to recycle 90 percent of all plastic bottles by 2029.

ISABEL WIJSEN   MELATI WIJSEN

Bye Bye Plastic Bags is helping reduce plastic waste in over 25 locations.

    Individuals are trying to help too. In 2013, two sisters from Bali, Indonesia, decided to take action. Isabel and Melati Wijsen were only 10 and 12 years old when they decided to take on plastic waste. They got the attention of the Bali government, which started creating its own bans and campaigns against plastic. The sisters also started Bye Bye Plastic Bags, a youth organization helping reduce, or cut down on, plastic waste around the world.

23

CHAPTER ELEVEN

INVENTOR BOYAN SLAT

In 2011, a Dutch teenager named Boyan Slat was swimming off the coast of Greece when he saw more plastic bags than fish. Upset, he decided to invent something to help. Throughout high school and college, Slat worked on ideas. He created a floating system to collect ocean plastics and remove them without harming ocean animals.

Boyan Slat also created a river barge, or boat, called the Interceptor. One Interceptor can keep about 55 tons (50 mt) of garbage from entering the oceans each day.

As Slat got older, he kept working on the system. In 2013, he started an organization called The Ocean Cleanup. It makes **technologies** to remove plastic from the oceans.

In 2016, Slat used his first system in the ocean near Japan. It didn't work well, but Slat and his team created a new system. In 2019, the cleanup system started working. The Ocean Cleanup hopes to remove 90 percent of plastic waste from the world's oceans by 2040.

CHAPTER TWELVE

NEW MATERIALS, NEW ANSWERS

There are many people working to find answers for the plastic problem. People have created new materials as alternatives, or other possibilities, to use instead of plastics.

One new kind of plastic isn't made from fossil fuels. It's made from sugar and carbon dioxide, and it's biodegradable. It can break down with **enzymes** found in soil bacteria.

Plastic water bottles create a lot of waste. A **sustainable** packaging company has created a new package material called Notpla. Notpla is made from seaweed and other plants. It only takes a few weeks to biodegrade. The material is also the company's name. Notpla also created clear bubbles called Ooho that can store water and are safe to eat.

These new materials might be good answers to the world's plastic bottle problem.

You can eat Ooho plastic water bubbles! They're also safe for the environment.

CHAPTER THIRTEEN

HOW YOU CAN HELP

Everyone can help solve the plastic problem. Here are some ideas to get you and your friends started:

- Start a plastic bottle collection club. In many states, plastic bottles can be returned to stores for money. You can help clean up where you live and get some money too!

- With so much plastic lying around, why not use it to make something new? Making things into something new can be fun. Use plastic trash to make art, planters, magazine racks, or baskets. Remember to use only plastic trash, not new plastic. The goal is to clean up.

- Start a home recycling station. Many things need to be sorted before they can be recycled. Use cardboard boxes to sort metal, plastic, cardboard, and glass. Label the boxes so everyone knows exactly what goes where.

This greenhouse was made from wood and plastic bottles. Can you think of something to make with the plastic you find?

CHAPTER FOURTEEN

A NEW WAY FORWARD

Even though plastic waste is a problem, some companies are working hard to fix it.

In 2018, the toy company LEGO created plant-based plastic bricks. People can easily recycle these building bricks. The same year, LEGO also recycled 93 percent of its trash. By 2030, LEGO will make all its bricks from sustainable materials.

A company called Agilyx has created a new way to recycle plastic using chemicals. It can even recycle single-use items that are usually thrown away.

Because plastic is so strong, a company called MacRebur is reusing plastic to make roads. It's used its new plastic mixture to build roads in England and San Diego, California.

Ideas can come from anyone, including you! No idea is too small and every effort helps. Get started today!

GLOSSARY

celluloid (SEL-yuh-loid) A plastic made mainly from plant-based materials.

contaminate (kuhn-TAA-muh-nayt) To pollute.

Earth Day (URTH DAY) April 22, a day set aside around the world to help promote environmental awareness for planet Earth.

environment (ihn-VIY-ruhn-muht) The natural world around us.

enzyme (EHN-zym) A chemical substance that helps natural processes in plants and animals.

fossil fuel (FAH-suhl FYOOL) A fuel—such as coal, oil, or natural gas—that is formed in the earth from dead plants or animals.

Great Pacific Garbage Patch (GRAYT puh-SIH-fik GAHR-bij PACH) A region of circular currents located in the northern Pacific Ocean that captures ocean waste.

material (muh-TEER-ee-uhl) Something from which something else can be made.

sustainable (suh-STAY-nuh-buhl) Able to last a long time.

technology (tek-NAH-luh-jee) A method that uses science to solve problems and the tools used to solve those problems.

tusk (TUSK) A long, large pointed tooth that comes out of the mouth of an animal.

INDEX

B
Baekeland, Leo, 6, 7
Bakelite, 7

C
celluloid, 6
chemicals, 7, 17, 30
China, 22
Clean Air Act, 12
coal, 4, 7
Congress, U.S., 12

E
Earth Day, 10
Environmental Protection Agency (EPA), 12
European Parliament, 22

F
fossil fuels, 4, 26

G
Great Pacific Garbage Patch, 14, 18, 19

H
Hyatt, John Wesley, 6

K
Keep America Beautiful, 12

L
landfills, 10, 17
LEGO, 30

M
Mestral, George de, 8
microplastics, 17, 19
Moore, Charles, 14

N
Nixon, Richard, 12
Notpla, 26
nylon, 7, 8, 9

O
Ocean Cleanup, 25
Ooho, 26

P
plastic bags, 10, 14, 15, 20, 21, 23, 24

S
single-use items, 14, 20, 22, 30
Slat, Boyan, 24, 25
Solid Waste Disposal Act, 12

W
Wijsen, Isabel and Melati, 23
World War II, 8

PRIMARY SOURCE LIST

Page 9
Women standing in line for nylon stockings at Miller's Department Store. Photograph. Ed Westcott. Oak Ridge, Tennessee. January 4, 1946.

Page 11
Earth Day poster. Designed by Yukihisa Isobe. 1970. Now kept at the Library of Congress.

Page 13
Recycling poster. 1970. Now kept at the Library of Congress.

WEBSITES

Due to the changing nature of Internet links, PowerKids Press has developed an online list of websites related to the subject of this book. This site is updated regularly. Please use this link to access the list: www.powerkidslinks.com/SOOF/earthsplastic